1 MONTH OF
FREE
READING

at

www.ForgottenBooks.com

By purchasing this book you are eligible for one month membership to ForgottenBooks.com, giving you unlimited access to our entire collection of over 1,000,000 titles via our web site and mobile apps.

To claim your free month visit:
www.forgottenbooks.com/free1193529

ISBN 978-0-331-48852-4
PIBN 11193529

GEOLOGICAL REPORT

Olinda Allen

OF

LANDS

BELONGING TO

𝕿𝖍𝖊 𝕽𝖎𝖉𝖌𝖜𝖆𝖞 𝕮𝖔𝖒𝖕𝖆𝖓𝖞,

IN

ELK COUNTY, PENNSYLVANIA.

BY

Dr. CHARLES T. JACKSON,

GEOLOGIST AND ASSAYER TO THE STATE OF MASSACHUSETTS.

PHILADELPHIA:

PARRY & McMILLAN,

1856.

GEOLOGICAL REPORT

OF

LANDS

BELONGING TO

THE RIDGWAY COMPANY,

IN

ELK COUNTY, PENNSYLVANIA.

BY

Dr. CHARLES T. JACKSON,

GEOLOGIST AND ASSAYER TO THE STATE OF MASSACHUSETTS.

PHILADELPHIA:

PARRY & McMILLAN,

1856.

STATE ASSAYER'S OFFICE,

33 Somerset Street,

Boston, July 24th, 1856.

Chas. K. Landis, Esq.
President of the Ridgway Farm and Land Company.

Dear Sir,

During the past month, I have at your request examined the geological character and resources of the Ridgway Farm and Land Company's property in Elk County, Pennsylvania, and have now the honor of presenting to you my report.

The tract of land in question comprises some twenty-seven thousand acres, and is eligibly situated, having a temperate climate, and being perfectly healthy.

The latitude of the Village of Saint Mary's is 41° 25' North, and its longitude is 1° 40' West of the City of Washington. Its elevation above the Sea is about 1,600 feet, as near as can be estimated from the range of the barometer.

It is about one hundred miles East of Lake Erie, and the tract of the Sunbury and Erie Railroad to the City of Erie, is about one hundred and ten miles.

The country is not mountainous, but is gently undulating, none of the hills being too steep or high to interfere with easy culture. The forest land is most heavily timbered with gigantic Hemlock, Curly Maple, Bass, Beach, Ash, Black Birch, Chestnut, Oak, Pine, and Cherry Trees, which by their thrifty condition amply prove the luxurience of the soil.

Some of the Chestnut trees measured by me were 13½ and 16½ feet in circumference, and some of the Pines were 13 feet 2 inches in circumference, and 137 feet high; an Ash tree measured 10 feet 3 inches in circumference, and the Hemlocks average at least 8 feet around the butt. The Maple and Cherry trees are from 6 to 8 feet in circumference and are very sound and perfect. At the saw mills I measured many White Pine logs that were from 3 feet to 3 feet 11 inches in diameter, and the Cherry logs were from 2 feet to 2¼ feet in diameter, and very clear from knots, as was evident on inspection of the sawn boards.

The Hemlock logs were from 3 to 4 feet in diameter, and those of Bass Wood generally were three feet, and the Ash were from 2½ to 3 feet, and those of Curly Maple 3 feet at the butt.

By this examination of the timber actually at the saw mills, some idea may be formed of the workable lumber cut and brought out from the contiguous forests. It appears that the forest timber is not only unusually large, but it is also remarkably sound and very clear from knots. The Trees generally on the Company's lands run remarkably high before throwing out branches, so that it is not uncommon to find clean trunks 80 or more feet in length.

The best soil appears to be indicated by the great abundance of Maple, Black Birch and Ash trees, which abound in the northern portions of the forest land; but little of this has been cleared and cultivated.—/ Owing to the abundance of Hemlock Trees near St. Marys, it has been thought that large Tanneries might be advantageously established in that town, and I have no doubt of the feasibility of this project, which will follow quickly after the completion of the Sunbury

and Erie Railroad, which will establish a ready communication with the markets, and allow the hides to
be carried to the tanneries at little cost. It is always
advisable to carry the hides to the tan bark, since they
are of so much less bulk and weight than the bark itself.
The moment an adequate market is afforded by the completion of the above mentioned Railroad, a large in-
come will be derived from the tan bark, as it is of an
excellent quality, and abounds in great quantities.—
The character of the soil of this region of country, as
proved not only by its geology, but also by practical
experiments, is well adapted to the growth of the Cereal
Grains and Grasses; Rye, Oats, Barley and Wheat
thrive very well, but Rye and Oats I observed were
most extensively cultivated and were very thrifty.
The last Winter having been one of unusual severity,
had injured the crop of Winter Wheat considerably,
so that the crops did not appear to stand even and re-
gularly in the fields, but I was informed by the farmers
that their wheat was very rarely winter-killed or
thrown out by the frost. From some cause not learned
by me, Corn was not extensively raised at St. Mary's.
I suppose that the small grains are preferred by the
people, who are nearly all Germans, and have been only
a few years in this country. If there is any difficulty
in the ripening of Corn, this may be easily remedied
by the selection of early northern varieties such as are
cultivated in Massachusetts and New Hampshire.—
From the samples I noticed at the Grist Mills, I saw
that the last year's Corn was not thoroughly ripened,
and that it was a Southern variety and not adapted to
the climate of this district.

As a grazing country nothing can surpass the Company's lands, for Clover, Herdsgrass and Red-top thrive

in the most luxuriant manner, and yield heavy crops of the best kinds of hay, and furnish the richest pastures. Cattle will be raised in large numbers when the Railway shall open a ready market for beef, and large profits will be derived, especially from the raising of Sheep. All our usual garden vegetables grow luxuriantly in St. Mary's, and the kitchen gardens bear testimony not only to the goodness of the soil, but also to the skill of the German women, who generally take the entire care of them, and while they raise esculents for the table, do not forget to cultivate also an abundance of beautiful flowers, with which to decorate their parlors. The uniform testimony of all the farmers of St. Mary's, I found to be in favor of the goodness of the soil.

The soil is a good mellow loam, sufficiently retentive of manures and of moisture, without being clayey or stoney. Its geological origin is from sandstone, limestone, shales, and fire-clays of the coal formation, materials well known to produce good soils when sufficiently supplied with mould, as is the case in the Company's lands. Judging from the state of the crops at the time of my visit, and from the blooming of wild wood flowers, I should regard the climate as like that of Connecticut and Massachusetts, the elevation of the land being nearly equivalent to a degree of latitude toward the North for temperature.

The population of St. Mary's alone now numbers about 2,500 persons. They are mostly Germans, except those who have recently settled, and who are becoming quite numerous; they are usually farmers from Pennsylvania, New York, New Jersey, and New England. The inhabitants appear to be a happy and contented people, having no political or religious dissensions among themselves.

--- ◂◦◦▸ ---

GEOLOGICAL RESOURCES.

The lands of the Ridgway Farm and Land Company are all in the carboniferous group, and contain the usual rocks of the bituminous coal field, namely : grey sand-stones, shales, limestones, and fire clay, with regular beds of good bituminous coal. These strata are nearly horizontal, not in any case dipping more than five degrees from the horizon, and generally between two and three degrees.

The sandstone forms a good building material, and will serve for the construction of Iron Furnaces, which will be lined with bricks made of the fire-clay, which is found beneath each bed of coal.

The limestone will serve not only for building mortar and agriculture, but also for fluxing the iron ores in the process of smelting iron. The fire-clay will serve for making fire-proof bricks, both for home use and for exportation, of which a large business is already carried on in the adjoining county of Clinton.— Iron Ores, namely : carbonate of iron, in large balls and masses, and brown hæmatite occur, with the shales and fire-clays of the coal strata, and may be obtained in adequate quantities for supplying blast furnaces, which will be erected when the coal mines shall have gone into operation.

Numerous beds of excellent bituminous coal, with a stratum of slaty Cannel over them, occur in very convenient locations for ready mining, and will be most extensively wrought when the western division of the Sunbury and Erie Railroad is finished, so as to open a direct communication between St. Mary's and Lake Erie; this division is now in active process of completion. An immense trade in coal and iron will spring up in this district in the course of a few years, and a great portion of the large demand for the lake navigation will be chiefly supplied from the Company's lands, for there is no other conveniently situated place, where a large supply of coals can be obtained so accessible by means of Railways.

At the present time I learn that the demand for coals upon the great lakes is far beyond the supply that can now be obtained, and this demand is rapidly increasing, owing to the augmentation of steam navigation on the lakes. Furnaces for smelting iron and other ores, and numerous founderies will spring up on the lake shore, when an adequate supply of good coals can be obtained at reasonable cost.

Few of our citizens upon the coast realize the extent and importance of the trade upon the great lakes, and many will be surprised to learn that the tonnage of the lakes considerably exceeds that of the sea coast of the United States, but such is undeniably the case. Steam navigation is destined to increase on the lakes in a far greater ratio, and the Canadas will not fail to avail themselves of the coals from the great bituminous coal fields of western Pennsylvania.

It will be seen by the results of my chemical analyses of the coals, that they are admirably adapted for the making of illuminating gas, and also for steamboat and furnace uses, besides being excellent for warming houses, and for all uses to which the best bituminous coals are applied.

The slaty Cannel coals will serve for the manufacture of the bituminous oil which has lately come into use for illumination, and for oiling machinery. The cannel coal will also serve for making gas, but its coke is not so valuable as that from other coals on account of the larger proportion of ashes it yields when burned. The coals from the principal beds make excellent coke, quite compact and suitable for use in smelting furnaces, for reduction of iron from the ores, and also for founderies where iron is to be melted and cast.

From the observations I have made, I am of opinion that there are at least *six beds of Coal* in the Company's lands, and but few of the upper ones only have suffered any considerable loss by denudation, or washing away by the formation of runs and valleys of excavation by the action of water.

It is impossible without sinking a shaft, or boring through the strata, to establish the order of position of the several beds with certainty; since the openings now made are too limited to admit of accuracy in determination of their dip or inclination by which their positions could be made known when the levels of the different points were also measured.

Allowing that there are but two workable beds of coal, one four, and the other six feet in thickness, and that for every foot of thickness of a coal bed, 1000 tons of coal is to be estimated per acre, (the admitted rule,) then there will be no less than 10,000 tons of coal in each acre of land underlaid by these two beds.

In many locations in the Company's lands, there are two four feet beds, and one of six feet in thickness.

I do not estimate any bed less than three feet in thickness as a workable bed, though I am aware that two feet beds are sometimes worked to advantage **when** the coals happen to be of good quality, and **are** favorably situated as to ready drainage.

There will be no necessity for working in **any** of the thin beds for ages to come, as there is such an extensive area underlaid by larger veins.

I shall give an account of the outcrops examined by me, and by consulting the map of your lands, you will perceive that the coal is proved to exist under **the** village of St. Mary's, and all the land immediately around that place.

There are also numerous exposures of out crops of coal beds on the road-side, and in the banks of small streams, or runs as they are called; their importance is not yet developed by mining operations, so it is impossible to estimate them at present; those here described are beds that have been sufficiently opened **to** demonstrate the quality of the coals and their thickness.

Dr. C. R. Earley's Coal bed exists south-west of Centreville, on the bank of a small stream called Toby's run, where it crops out and shows itself for a considerable distance in the rocky bank on the brook. The strata, including the coals, dip to the north-westward 5°, and the course of the out crop is N. 30° E.—S. 30° W. The principal bed is 3 feet 9 inches thick; the coal is of good quality; this is underlined by about 20 inches of fire-clay, and is overlaid by slaty cannel coal and bituminous shales. A bed of buff-coloured limestone underlies this coal, and is seen in the brook below— another bed of coal about 20 inches thick exists above the principal bed, but since it cannot be worked in the same drift with the workable bed, we regard it as **of** little importance. Not far from this place we find the coal mines opened by Jonathan Keller; this bed is not opened down to the fire-clay, and I could only measure a thickness of coal of 3 feet 4 inches, though Mr. Keller says that it is 4 feet 4 inches thick. There are said to be two beds of coal above this, and one below, but they are not in a condition to admit of examination, the earth having caved in and filled up the openings.

JOSEPH NEIST'S MINE

Is north of the village of St. Mary's, and quite near to it; this bed has been worked to some extent, and furnishes coal for the neighboring smitheries.

Mr. Neist works in the extraction of coals in the winter months, when he has little else to do, and sells his coal for 3 cents a bushel cash at the mines, or for 4 cents in barter.

This coal is underlaid by a good fire-clay and overlaid by shales containing balls of iron ore. The coal is 3 feet 9 inches thick, and dips 3° to the N. 70° W. A drift or level has been driven into this coal for the distance of 55 feet, and on the course of the dip, so that it is difficult to keep the water from collecting in the drift at its further extremity. This erroneous method of opening the coal mines in the vicinity of St. Mary's, is generally pursued by the people who, finding the out crops of the beds, follow down upon them without any systematic views to future workings. I shall have occasion to show that nearly all the mines should be opened on the opposite sides of the hills, where the out crops do not appear at the surface.

MACREADY'S MINE

Is south of the village of St. Mary's, and nearly half way between that village and Centreville. This bed is 3 feet 10 inches thick, and dips to the north-west-ward 5°. It rests on a bed of fire clay and is overlaid by slaty cannel coal and shales, containing balls of hæmatite iron ore and carbonate of iron. It is under-laid at some distance below by 9 feet or more of buff-colored limestone, which is exposed on the sides and in the bottom of a small brook, near the mine. Mr. Macready has cut four drifts into the coal, the prin-cipal one being 200 feet long. The Coals are sent in winter to the neighboring towns for sale.

JACOB TAYLOR'S MINE

Is situated quite near the village of Centreville, and is one of the most valuable in the district. This bed is not less than 6 feet 2 inches thick, but since it was not opened quite to the under-clay I could not measure its entire thickness. Mr. Taylor says it is seven feet thick. It is certainly in the most important bed of coal yet discovered in Elk County. Samples of this Coal proved on analysis to be of the best quality.

A. HOWES' MINE.

This is situated a little north of Centreville village; the coal is 5 feet 6 inches thick, and is of good quality. On the road-side we also saw the washed outcrops of a coal bed; and in a maple grove, called Kersey Sugar Bush, near the residence of John Sullivan, we found the coal outcropping on the side of a run or stream.

THERESA STREET MINE.

This is situated in St. Mary's. This bed is only 1 foot 8 inches thick, and is nearly horizontal. Iron ore in balls and flattened oval shaped masses also abounds, and this opening will prove valuable, chiefly on account of the abundance of the iron ores that can be obtained from it. The broken slaty rocks are compactly cemented by oxide of iron, which forms with them a firm hard pan.

GLOLT'S MINES.

No. 1 is a bed 2 feet 10 inches thick, with about 8 inches of slaty cannel coal on top. No. 2 is a bed 3 feet 11 inches thick. Both these beds rest on fire-clay. They are separated by too thick a mass of strata to admit of being wrought by a single line of levels and chambers. The lowest bed being thickest will be most advantageous to work.

BUCHEST'S MINE.

The Coal of this bed is 3 feet 9 inches thick, the measurement being taken 50 feet in and upon the level or drift. It is 3 feet 5 inches thick at the outcrop. Good coals are obtained at this mine, a level having been driven towards the north some 50 feet to expose the coal under cover, where it is best. Hernhauser's old bank of coal presents a bed of 3 feet thick.

In Paul Street, in St. Mary's, a bed of coal 2 feet 6 inches thick is exposed, and has with it iron ores in the shales, and fire-clay.

Anthony's Coal Mine is in a bed 3 feet 11 inches thick, with a thin bed of slaty cannel coal on top, and then over that 6 feet of bituminous shales, filled with nodules of hæmatite iron ore.

THE PRIST MINE

Has been opened in a bed of coal 3 feet 7 inches in thickness; a drift of 30 feet in extent being driven into the hill, in a course S. 15° E. The coals are of good quality.

In the Sugar Grove we also found two beds of coal, one 2 feet 4 inches thick, and the other was not opened so as to admit of measurement. The dip of the coal was N. W. 5°.

Now on looking upon the plan of the company's lands, and noting the courses of the different coal out-crops, which are at right angles with the dips, *it will be found impossible to reduce the number of beds of coal below five*, even taking into account the differences of level of the mines opened

We shall however, make the number and order of succession known if we bore perpendicularly through all the strata at some convenient point, and this I advise to have done previous to sinking shafts and establishing regular mining works. Since the coals are most frequently exposed on the south east sides of hills and ravines, and the general dip of the coal is to the N. W. it is obvious that the coal will be under good cover of rocks on the N. W. side of the moderate eleva-tions of this district, and that the coal should then be sought by sinking shafts, and the levels or gangways will be driven up the slope of the strata, while the water running to sump at the bottom of the engine shaft will be pumped out, and the mine will thus be kept dry.

In some cases it will be practicable to reach the coals by drifts driven horizontally in the northwest sides of the hills, but this will only be in rare instances.

The present method of sinking on the outcrops should be at once abandoned, as injurious to the future working of the mines, by letting in the water.

Large quantities of excellent iron ore will be raised in working the coal mines, and this should be laid aside in heaps and be kept for sale to the iron furnaces that will ere long be established. In the ploughed fields nearly every stone seen is a mass of iron ore, containing from 40 to 50 per cent. of iron. These should also be collected for the furnaces.

The value of the grey sandstone as building material, has been sufficiently proved by its use in building several excellent dwelling houses, and especially in the construction of the large Church of Saint Mary's, which is entirely built of this stone in its rough state. The more compact strata of this rock will serve for hearth-stones and tymps to the smelting furnaces.

CHEMICAL EXAMINATION OF THE COALS

FROM THE MINES IN RIDGWAY FARM AND LAND
COMPANY'S PROPERTY, ELK COUNTY, PA.

Examination of their value for Gas-making.

No. 1.

Specimen from Taylor's 6 feet 2 inches bed.

100 grains gave Coke,　60
Gas, - - - - - - 40
───
100

No. 2.

Slaty Cannel, ordinary kind.

Uncemented Coke,　76
Gas, - - - - - 24
───
100

No. 3.

Bituminous Coal—Macready's Mine.

Good Coke,　61
Gas,　- - 39
───
100

No. 4.

Bituminous Coal, out-crop yielded
Coke, - - - - 63 2
Gas, - - - - 36 8
───
100 0

CHEMICAL ANALYSIS OF A SAMPLE OF THE BEST COAL.

One hundred grains yield

Fixed Carbon, - - - - - -	52 38
Gas driven off by red heat, -	40 00
Grey Ashes, - - - - - - -	7 62
	100 00

The Ashes analysed yielded

Silica, - - - - - - - - -	6 22
Oxide of Iron and Alumina, -	1 10
Lime, - - - - - - - - -	0 22
	7 54

The Slaty Cannel Coal yielded

Fixed Carbon, - -	32
Gas, - - - - - -	24
Earthy Matter, - -	44

ANALYSIS OF THE BUFF COLORED LIMESTONE.

One hundred grains yielded

Carbonate of Lime, - - - -	95 75
Silex, - - - - - - - - -	3 00
Oxide of Iron, - - - - - -	1 25
	100 00

Lime in 95 75 Carbonate of Lime, $53\frac{73}{100}$

CHEMICAL ANALYSIS OF THE WHITE CAR-BONATE OF IRON (IRON ORE BALLS,)

OF THE RIDGWAY FARM AND LAND COMPANY'S MINES.

100 grains of this ore yield

Prot-Oxide of Iron,	- - - -	61 50	
Carbonic Acid Gas,	- - - -	31 50	diff.
Silex, - - - - - - - - - -		7 00	

Metallic Iron in 61 50 Prot-Ox.—$47\frac{84}{100}$ } 100 00

The loss by roasting the ore at a red heat is but 24.75, and the prot-oxide of the white carbonate becomes per-oxide of iron.

By roasting in kilns in the large way, probably the ore will be enriched about 20 per cent. by removal of a portion of the carbonic acid, and the moisture.

Then the ore will produce more than 51 per cent. of metallic iron.

I found no sulphur or other impurites in this ore that would prove injurious to the iron, but now and then balls will be found which contain sulphur, and these should be rejected from the heap. Haematite iron ores, such as are found in the shales, and occur loose in the soil, yield after having been properly roasted, about fifty per cent. of iron on reduction in the furnace.

In conclusion, I would assure you and the gentlemen associated with you, that I entertain a very high opinion of this valuable tract of land; valuable alike for timber and a productive soil, but still more valuable for its immense mineral resources in iron ores and coals—the great material powers which advance civilization more rapidly than any other means. Iron and Coal have made England what she now is, and the time will come when the North-western Coal Basin of Pennsylvania will form one of the richest portions of that wealthy State.

Respectfully your Obedient Servant,

CHARLES T. JACKSON, M. D.

Assayer to the State of Massachusets, &c.

CPSIA information can be obtained
at www.ICGtesting.com
Printed in the USA
BVHW04*1422041018
529155BV00028B/490/P

9 780331 488524